Broodiness in Domestic Fowl
And its Inheritance in Rhode Island Reds

by Massachusetts Agricultural Experiment Station

with an introduction by Jackson Chambers

This work contains material that was originally published in 1920.

This publication is within the Public Domain.

This edition is reprinted for educational purposes
and in accordance with all applicable Federal Laws.

Introduction Copyright 2018 by Jackson Chambers

The World's Largest Selection of Vintage Poultry Books

www.VintagePoultry.com

Self Reliance Books

Get more historic titles on animal and stock breeding, gardening and old fashioned skills by visiting us at:

http://selfreliancebooks.blogspot.com/

Introduction

I am pleased to present yet another title on Poultry.

The work is in the Public Domain and is re-printed here in accordance with Federal Laws.

As with all reprinted books of this age that are intended to perfectly reproduce the original edition, considerable pains and effort had to be undertaken to correct fading and sometimes outright damage to existing proofs of this title. At times, this task is quite monumental, requiring an almost total "rebuilding" of some pages from digital proofs of multiple copies. Despite this, imperfections still sometimes exist in the final proof and may detract from the visual appearance of the text.

I hope you enjoy reading this book as much as I enjoyed making it available to readers again.

Jackson Chambers

CONTENTS.

	PAGE
Introduction,	93
The material and its treatment,	95
Recognition and treatment of broody birds,	95
Variation in amount of broodiness,	96
Germinal basis of broodiness,	102
Modifying factors for broodiness,	106
The production of a strain of low degree of broodiness through selection,	107
Relation between birds of a low degree of broodiness and absence of broodiness,	113
The interrelation of several broody characters,	114
Summary and conclusions,	116
Literature cited,	116

BULLETIN No. 199.

DEPARTMENT OF POULTRY HUSBANDRY.

BROODINESS IN DOMESTIC FOWL.

DATA CONCERNING ITS INHERITANCE IN THE RHODE ISLAND RED BREED.

BY H. D. GOODALE, RUBY SANBORN AND DONALD WHITE.

INTRODUCTION.

Broodiness, as pointed out by Herrick (1907a, 1907b), is one phase of a recurring cyclical process in birds. In the domestic fowl when kept primarily for egg production, the instinct is not allowed to run its normal course, but is checked by suitable means in its initial objective stages. Some individuals, however, never exhibit the instinct. In this study of the inheritance of broodiness two categories of birds may be recognized, viz., those that exhibit the initial stages of broodiness, which are promptly checked, and those that do not exhibit any signs of broodiness. Broodiness is intimately connected with egg production, and, other factors being equal, its presence or absence determines the number of eggs laid, since, as shown later, its presence tends towards decreased production. A knowledge of its inheritance should show the steps necessary for its complete elimination from a flock.

The character, moreover, is not a superficial and unimportant one, but is a well-defined characteristic of the class Aves, and is essential for the survival of every species in the class. If the instinct were lost in a state of nature, without being replaced by some compensating mechanism,[1] the race would become extinct. In nature selection is constantly directed in favor of the character, since those individuals that lack it will leave no progeny, yet among domestic fowl we find entire races in which the character is lacking.

Poultrymen recognize both broody and non-broody races. The American breeds, *i.e.*, Plymouth Rocks, Rhode Island Reds and Wyandottes, and

---

[1] It is well known, of course, that the American cowbird and the European cuckoo have developed a compensating mechanism.

the Asiatics, *i.e.*, Langshans, Cochins and Brahmas, may be cited as examples of the former, while Hamburgs and Campines, and the Mediterranean breeds, *i.e.*, Leghorns, Spanish and Anconas, furnish examples of the latter. The distinction is based on the proportion between broody and non-broody individuals in each race, for some non-broody individuals occur among the broody races, while records are lacking to show that broody individuals are entirely absent from any of the non-broody races. The Leghorns are commonly regarded as a non-broody race, but as shown in Table VI, taken from the report of the fifth laying contest at Storrs (Kirkpatrick and Card, 1917), a considerable number become broody. It is a matter of common knowledge among poultry keepers that among the broody races there are considerable differences, some races, of which the Rhode Island Reds are an example, having an intense development of broodiness compared with others, such as the Barred Plymouth Rocks, in which the amount of broodiness is relatively slight.

There are few published reports on the character in the domestic fowl, though there is, of course, a considerable amount of matter scattered through the poultry literature, in which broodiness is mentioned in a more or less general way, but which is of no importance from the standpoint of this paper. Both Bateson's (1902) and Hurst's (1905) data showed that in a cross between a broody and non-broody race, broodiness was dominant, but they have published no further observations. Pearl (1914) has published certain data relating to broodiness, with which in general our data agree. A repetition of the same sort of material is unnecessary here. His methods of collecting the data and of handling the broody birds also are essentially the same as our practices in these respects. In general, our experience with this instinct agrees with his, except that there are two points for which different interpretations may be presented. On page 285 (*loc. cit.*) he makes this statement: "It appears to be the case that in the domestic fowl the brooding instinct has to a very large degree disappeared along with the fact of domestication." Evidently this author had not encountered a strain like our Rhode Island Reds, for such a statement would be impossible after an experience with such a strain. In the second place, we entertain some doubt as to the advisability of measuring the intensity of broodiness by the length of the non-productive period associated with the objective symptoms of broodiness (*loc. cit.*, page 273), because, while the cessation of egg production coincides in nearly every instance with the onset of objective symptoms, the resumption of production is often delayed by other factors, among which may be noted the innate capacity for egg production and readiness to molt. In regard to its effect on egg production, Goodale (1918) makes the statement that the ratio between the egg production prior to the first broody period and that subsequent thereto is about 100:60.[1] Gerhartz (1914) has studied the metabolism of the broody hen in connection with his studies on the metabolism of the laying hen.

---

[1] The data on which this statement is based are given for the first time in Table VIII.

The Material and its Treatment.

The materials for the present study of the inheritance of broodiness are the pullet-year trap-nest records of the flock of Rhode Island Reds, bred at this station from 1913 to 1917, primarily to furnish data on the inheritance of fecundity. The usefulness of these data is limited in one important respect, since, as discussed in the section on variation, a year's record is not long enough to determine a hen's capacity for becoming broody. Limitations in housing capacity and labor have hitherto prevented the retention of non-broodies as long as was desirable. In the handling of the data, therefore, we have classified birds as broody or non-broody on the basis of the pullet-year records only, even though on this basis the non-broody class will contain more birds than it should. However, the theories of the inheritance of broodiness to which we have been led could not be substantiated from the available data, even if the difficulty under discussion were removed.

Recognition and Treatment of Broody Birds.

The recognition of a broody period is an easy matter with slight experience. The onset of broodiness is usually sudden. On the last visit to the trap nests late in the afternoon one or more birds are found that are very much disinclined to leave the nest. If they cluck and ruffle their feathers the diagnosis is certain, and the birds are removed to the broody coop to be "broken up." Sometimes part of the symptoms are lacking. In case of doubt the bird is merely removed from the nest. By the following afternoon, if she is really broody, all symptoms are well manifest. Mistakes are not easily made.

The broody coop in which the broody hens are confined, in order to prevent the instinct from running its normal course, is a box with slatted sides, top and bottom. The routine practice in dealing with broodies is to place all the broodies found in each pen in one of these coops. The same coop also receives the broodies on each of the two days following. Three days later the entire lot is released as a unit. Thus, the birds are confined from three to six days each, a period which is sufficient for the majority to "recover from the attack." A few, however, require a longer period of confinement. The confined broody individuals are supplied *ad libitum* with the same sort of food and water supplied the rest of the flock.

A bird must, as a rule, be classified either as broody or not broody, though in a few rare instances birds have exhibited a part of the broody symptoms only, as, for example, when a hen clucks and ruffles her feathers, but does not remain on the nest continuously, nor cease laying.

Variation in Amount of Broodiness.

A bird once broody may exhibit the character in various degrees which can be classified under two heads, — first, variation in the number of times a bird becomes broody in a period of given duration, usually the laying year; and second, variation in length and intensity of the individual broody periods. The latter is the less important of the two, for the length of time required by the vast majority of birds to "recover" from the attack is of comparatively small importance. To be sure, some birds take double the time that others require in "recovering," but it is an extreme case when more than a week is required, if no extraneous factors are present. Further, on forming a correlation table between number of days confined and subsequent egg production, it became evident that the coefficient of correlation (though not calculated) was so small that no relationship of importance existed between the subject and relative. The large factor in variability in broodiness is found in the variation in the number of times the broody cycle is repeated, as is shown later on.

Four sample egg records are shown here to illustrate individual variations in amount of broodiness. A numeral in a square indicates the hour at which an egg was collected; B. L., broody and placed in broody coop; A, released from broody coop; N, associated with a numeral, means that the bird visited the nest, but did not lay.

No. 8314
PEN

DATE	1	2	3	4	5	6	7	8	9	10	11	12	13	14	15	16	17	18	19	20	21	22	23	24	25	26	27	28	29	30	31	TOTALS	
SEPT.																																	
OCT.																			12	10			10.5		8	1	2.5		8	10.5	11.5	9	
NOV.	1	4		8	10.5	1		1	1	3	11.5	12 BL			A												N9	1	N1	7.5		12	
DEC.	4.5		7.5	1	N3.5	7.5	11.5	3		9	10.5	1	4		9	1		9	3		10.5	3		4.5		BL						18	
JAN.	A												N3	N1	N11.6 1		10.5	3		10.5		1		10.5	3		10.5	1		9	1	12	
FEB.	4			10.5	3		10.5	3		10.5	3 N4.5	BL		A												N4		10.5	3			9	
																																	60
MAR.		8.5	10.5	2		9	10.5	10.5	11.5	1	12	1	2	3		7	10.5 BL				A	N5		9	1		9	N1	2		7.5	19	
APR.	8.5	9.5	10	10	10	10	10.5	1 BL		A BL						A	10.5	2		7.5	9	9	8.5	8.5	N3	9x	10	10	2	BL		19	
MAY						A		1		7.5	11									10.5	9			10				9		2 N3	BL	8	
JUNE						A							11	10	1	1	1	1	2.5		7.5	1 BL					A					9	
JULY				9	11.5	2	3		7.5	11	1		8 BL					A						9	3		1	11	11	1		14	
AUG.										1		9	11	3		9	9	11	11	1		BL								A		9	
SEPT.						10	2		8.5	11	11	1	12	1	3 BL																	9	
OCT.		1		9	3		9	10	11	11	11	1		1	BL																	10	
NOV.																																	

YEAR'S TOTAL 157
(365 DAYS.)

FIG. 1.—Egg record of a very broody individual.

NO. B3700

PEN

DATE	1	2	3	4	5	6	7	8	9	10	11	12	13	14	15	16	17	18	19	20	21	22	23	24	25	26	27	28	29	30	31	TOTALS	
SEPT.																																	
OCT.																																	
NOV.																					11.5		7	9.5	3		9	2		8		7	
DEC.	N3	N3.5	3		7.5	9.5	2.5		7.5	8	10	11	11.5	3			9	10	1	4			3		8	9.5	11	4		7.5	10	21	
JAN.	11.5	N3.5		7.5	8.5	10	11.5	4.5		8	9.5	10.5	2	3.5		8	11	11.5	1	4		8	9.5	11	10.5	10.5	11	N2.5	N4.5		8	23	
FEB.	11.5	12.5	4		8	10	11	N3		4		10.5	10	11.5	3.5			8.5	10	11.5	3.5		8.5	8.5	10	1	1.5	4.5				21	
																																	72
MAR.		8	2		8.5	10	N3	N3.5		8.5	1.5		8.5	10	1.5		8	9.5	12	11.5	11.5	11.5	N2	4 N5.5		8.5	10.5	4		N1	11.5	20	
APR.	3	3	N4	4		9.5	9.5	N2	N6		9.5	11	N2.5			10	11.5	2	12.5	1	2.5	4.5		N1	9.5	11	12	1.5 N5.5	BL			18	
MAY.				A		N11		N5		9.5	N3	N5.5		9	9	9.5	9.5	10.5	10.5	11.5	10	N3	11.5	10	9	2 N5.5	BL					13	
JUNE			A				N3.5	4.5		9.5	11.5	11.5	11.5	12.5	3		9.5	9.5	11	11	11	3 N5.5	BL				A	N4.5				13	
JULY		N9							2		9.5	2.5		11	12	3.5		9	10	9.5	9 N5.5	1.5C BL				A						11	
AUG.									N9	12			9.5	9.5	10.5	N3.5	4.5			1.5		8.5	10	10	10	1.5 N5	N2 BL					11	
SEPT.				A														N9.5	3		10	11	5		N5	3		9.5	9.5	11		8	
OCT.	11	11	11	1	12.5	3 N5	BL						A																			6	
NOV.																																	

YEAR'S TOTAL 172
(365 DAYS)

FIG. 2. —Egg record of a bird becoming broody about midseason.

NO. B 3226
PEN

DATE	1	2	3	4	5	6	7	8	9	10	11	12	13	14	15	16	17	18	19	20	21	22	23	24	25	26	27	28	29	30	31	TOTALS
SEPT.																																
OCT.																																
NOV.									N3												N16 M15			8.5				7.5	8.5	11		7
DEC.	11	11	1.5	3		8.5	10	9.5	10.5	1.5	8	8	10	2		8	9	10	11	11	2	1	4	9	3	N4 8.5		10.5	3			24
JAN.	8.5	10	1.5	3.5		8	10	11.5	4		9.5	10.5	2	4.5		8	9.5	10	1	4	8	N1 11	11	3		8	9	9	11	11		25
FEB.	1	4.5					8	9	10	11	11.5	2.5	N3 10	1	3.5		8.5	1		2	8	8	10.5	3.5	3	8.5	11.5	3.5				22
MAR.	8	10.5	10.5	1.5		8	10.5	11.5	N3				8.5	N3 3.5			N1 2			7.5	10	11.5	11.5	12		1.5	1	1.5	6	5.5	9.5	21
APR.	12	12	2	2	6		9.5	11.5	2.5	2	5.5		9.5	11.5	12.5	1.5		9.5	11	2.5	4.5		9.5	11	12	1	11	12	12.5	5		25
MAY	9.5	N12		3	3.5	N6	12.5 N6	4.5 N5.5	BL					A N6.5				9.5		N5 N6		9	10.5 12		N4 3.5	6	9.5	9.5		12
JUNE	11.5	2.5				12.5	4.5		11.5	3		5	N1 15	1		N3.6 12.5	11.5	3	N3 N4 12.5 N2	3.5		N8 9.5			N4 N6 12	9.5	N5 N4 4.67	N5 N6 10.5	N9 2		19
JULY	9	11.5	3			9	N3 N5.5		11	12.5	5	10.5 1.5	5	10.5 1.5		5	1		11	3	11	3.5		8	2			9.5	2	9.5		22
AUG.	1	3.5		9	11.5 5.5	3	N3	10.5 3			10.5 12	4.5	N3	9.5 2	4.5	1.5		14.5		N3.5 N2.5	11.5 4	11.5 4	10			10	1.5		10.5	3.5	N3.5	20
SEPT.	N3 N3	3.5		10.5	3		11.5	5			N1 5	11	4.5		1.5																	11
OCT.	N11																															0
NOV.																																

YEAR'S TOTAL 208
(365 DAYS.)

Fig. 3.—Egg record of a bird becoming broody only once the first year, and that at midseason.

NO. B640
PEN

DATE	1	2	3	4	5	6	7	8	9	10	11	12	13	14	15	16	17	18	19	20	21	22	23	24	25	26	27	28	29	30	31	TOTALS
SEPT																														1		2
OCT.	8.5	10	4.5																								2					
NOV.	9	12		9	8	10	4.5		9	1.5	4		7.5	10	3		8	9	10	11	3		8	9	2	7	7.5	9.5	10			23
DEC.					11	10.5	3		7.5	9	1	3		9	10	11	4		7	9	9.5	11	1	2.5		7	10.5	3.5	7		11	23
JAN.	1	4	9	11	3		8	10	1.5	4		9	10.5	4		8.5	11	4		8	11	3.5		2.5	8.5	11	3		8.5	1.5	4	24
FEB.	10	1	4		11	3			8.5	2			8.5	12	4		10	3		8.5	3	4	9		11	4		9.5				20
MAR.	10	4		9.5	12	5		9.5			11	2	8	10	4		10	10			8.5	5.5		9	1.5		11	3		9.5	3	22
APR.	6				4			1	3		2.5		2		12	1	4			3		12.5	11		10	6		2		11		18
MAY	2	4			4		5			10	4			12		10	6		9.5	2		10	3	11	2	3	12.5	3	10.5	2	3	20
JUNE	1		10	12	3			2		12	3		11				4	11			10	3					1	3	1			17
JULY	5		3		11	W3 11	4.5			N10 3	11		11	4		11	5	5	11		1	1		11.5	4		1.5		NSMC 1	10		17
AUG.	1	5	11	3					10	3		11	3		10.2	5			11		5		10	10	1.5	10	10.5	10 3	5 NSMC BL	A	A	21
SEPT	BL	A																														0
OCT.																																0
NOV.																					10	10		10 1				10 1	10 1 1			5

YEAR'S TOTAL 207 (365 DAYS)

FIG. 4.—Egg record of a bird becoming broody at the end of the laying year.

Figures 1, 2 and 4 are examples of variability in number of broody cycles. It will be observed that there is a period of production of variable length before the first broody period makes its appearance. This may come soon after production begins (Fig. 1), or may be delayed till near the end of the season (Fig. 4), or even till subsequent seasons. After the first broody period subsequent periods occur in fairly regular sequence, the cycle of broody, rest and production being repeated over and over. It is obvious, then, that the number of broody periods occurring during the first year will be determined in part by the time of year the first broody period occurs. After one broody period it becomes a question of the number of additional cycles that are added before a bird stops laying for the season. In the extreme case of Fig. 4, production was not resumed after the first broody period, which came late in the season.

In later years there has been much less regularity in the recurrence of the broody periods in that part of the flock bred specifically for increased egg production, but the parallelism does not necessarily mean that the latter has caused the former. It is noticeable that the broody periods are fewer in number and limited to the height of the broody season in many individuals, which afterward may become regular producers, without further evidence of broodiness in that season. For example, instances are quite common where a single broody period occurs in mid-season, and is followed by continuous production, as shown in Fig. 3. Sometimes there are two or even three such periods followed by continuous production. Occasionally some birds have broody periods occurring at rather irregular intervals.

When birds are kept through the second season it is found that some birds that did not become broody the first season may become broody at some point in the second season. One instance occurred where the bird did not become broody till the third season. Because of the physical limitations in the matter of plant equipment we are unable to give the exact percentage of birds not broody in their pullet year that became broody later on. We have, of course, the data for such birds as were kept over, but we do not believe it gives a fair picture of what will be found in large flocks. They are, however, fairly numerous. There is, moreover, some evidence that various lines behave unlike others in this respect. On the other hand, only four instances free from complications have occurred among our records of birds that were broody the first season but failed to become broody thereafter. It is clear that the pullet year is a good index of the presence of the broody instinct for those birds that actually become broody, but not as good an index for those that do not become broody.

This brief description of variation in amount of broodiness, together with the data given in Table VIII, is sufficient to give the reader a general idea of its nature. Further details are outside the scope of this paper.

Germinal Basis of Broodiness.

The normal or wild condition is the presence of broodiness, otherwise the race could not survive. Hence, any changes from the normal condition of those parts of the germ plasm which are responsible for the existence (as distinguished from amount) of broodiness will probably result in a failure in the appearance of the broody instinct. Broodiness may also, of course, fail to become manifest from non-genetic causes, and hence some genetically broody birds are recorded as non-broody. Non-broodiness, therefore, is a comprehensive term employed to describe the phænotypic condition resulting from several sorts of genetic differences. The situation is parallel, in certain respects, to that of eye color in Drosophila. Red eye is the normal or wild color. Changes in the germ plasm result in a host of other eye colors, which are all non-red. In like manner, from the genetic standpoint, all broody birds are presumably alike in their fundamental broody mechanism, except for the homozygous or heterozygous condition, and, of course, the presence of modifiers discussed in the next paragraph. Non-broody birds, on the other hand, may belong to quite different genotypes. It is, therefore, improbable that any one scheme can be applied to the inheritance of non-broodiness in domestic fowl. Indeed, the data given in Tables II and III indicate clearly that several types of non-broodies exist in the Rhode Island Reds. The presence of these types suggests that still different genetic types of non-broodies will be discovered in other breeds.

It seems clear on general grounds that a distinction must be made between the primary factors concerned with broodiness and modifying factors. The latter may act in various ways on the primary mechanism for the manifestation of broodiness, but they cannot act unless that mechanism is present. We may expect that such modifying factors will control the intensity of broodiness in either a plus or minus direction, and, extending this reasoning to its logical conclusion, such modifying factors may prevent entirely any manifestation of broodiness. Non-broodiness, therefore, may result from some genetic change in the secondary mechanism, as well as in the primary.

There are three possible sorts of matings (the male being treated as though capable of giving phænotypic expression to his genotypic constitution), viz., both parents may be broody, both non-broody, or one broody and the other non-broody. From each of these three possible matings there are three possible groupings of the female offspring, *i.e.*, (1) all may be broody, (2) all non-broody, and (3) part broody, part non-broody, as shown in Table I.

TABLE I. — *Kinds of Offspring expected from all Possible Kinds of Matings.*

Parents.	Offspring.
Both parents broody, ...	All broody. All non-broody.* Part broody, part non-broody.
Both parents non-broody, ...	All broody.* All non-broody. Part broody, part non-broody.
One parent broody, the other non-broody, . . .	All broody. All non-broody.* Part broody, part non-broody.

Of the nine possible groupings between parents and offspring, six have been realized in our experience. Those marked with an asterisk (*) have not been realized. Two of the three unrealized possibilities should be realized eventually, and the third, all non-broody offspring from broody parents, is not expected, as is explained beyond.

The preceding table, as well as the ratios in which the offspring occur, Table II, does not agree with the assumption that broodiness is a simple Mendelian dominant and non-broodiness a simple recessive in all instances as Hurst (1905) supposed. If non-broodiness were a simple Mendelian recessive, then the son of a non-broody hen should throw either all non-broodies or half non-broodies when bred to non-broodies, but this does not always happen. Moreover, the establishment of a non-broody strain should have been a much simpler matter than it has proved.

TABLE II. — *A Comparison between Observed and Theoretical Ratios, assuming that Two Factors, A and C, must both be present in order that an Individual may become Broody.*

Father's Band Number.	1915.							1916.								1917.											
	Number of Mothers.			Offspring.					Number of Mothers.			Offspring.					Number of Mothers.			Offspring.							
				Observed Ratio.		Expected Ratio.						Observed Ratio.		Expected Ratio.						Observed Ratio.		Expected Ratio.					
	Non-broody.	Broody.	Total.	Non-broody.	Broody.	Non-broody.	Broody.	Father's Band Number.	Non-broody.	Broody.	Total.	Non-broody.	Broody.	Non-broody.	Broody.	Father's Band Number.	Non-broody.	Broody.	Total.	Non-broody.	Broody.	Non-broody.	Broody.				
68,	1	3	4	0	9	0	9	274	0	2	2	0	8	0	8	3003	4	1	5	23	1	22 1/8	1 7/8				
A143,	1	2	3	1	9	0	10	3617	0	5	5	6	24	3 1/2	26 1/2	5470	4	0	4	13	5	13	5				
228,	0	3	3	0	5	0	5	4312	0	1	1	0	8	0	8	5581	0	1	1	0	4	0	4				
269,	3	10	13	7	40	6	41	4354	1	4	5	5	32	2	35	7727	0	2	2	1	3	3/4	3 1/4				
270,	0	2	2	0	6	0	6	4378	0	4	4	6	22	5 1/2	22 1/2	7745	0	3	3	4	10	2 1/2	11 1/2				
271,	0	4	4	2	15	7/8	16 1/8	4723	0	6	6	15	59	16 7/8	57 1/8	7931	0	1	1	2	1	1 5/16	1 11/16				
272,	0	1	1	0	2	0	2	4786	0	2	2	2	6	7/8	7 1/8	8027	2	6	8	30	39	27 1/2	41 1/2				
274,	4	5	9	0	43	0	43	4882	0	1	1	5	20	6 1/4	18 3/4	8095	1	3	4	3	4	2 13/16	4 3/16				
A323,	0	1	1	0	1	0	1	5077	0	3	3	2	10	2 1/4	9 3/4	8097	0	5	5	21	22	17 1/4	25 3/4				
619,	0	2	2	0	3	0	3	5240	0	1	1	0	7	0	7	8147	2	5	7	15	33	18	30				
2313,	0	5	5	1	30	7/8	30 1/8	5470	0	1	1	1	2	1 1/2	1 1/2	8217	0	3	3	0	9	0	9				
2539,	0	1	1	0	1	0	1	5477	0	2	2	0	15	0	15	8422	1	2	3	7	4	6 1/2	4 1/2				
2914,	1	2	3	2	7	2 7/8	6 1/8	5581	0	7	7	2	42	1 1/2	42 1/2	8528	1	5	6	27	32	28 1/4	30 3/4				
3003,	4	1	5	11	3	11 3/8	2 5/8	5584	0	4	4	3	35	0	38	8761	0	4	4	9	15	7 5/16	16 11/16				
3533,	0	4	4	1	11	7/16	11 9/16	5642	1	0	1	0	2	0	2	9305	0	2	2	2	5	1 1/2	5 1/2				
3617,	1	10	11	1	38	1 1/4	37 3/4	5776	0	1	1	0	8	0	8	9752	6	5	11	57	16	48 3/8	24 5/8				
4128,	0	4	4	2	10	1 1/2	10 1/2	6265	0	3	3	1	19	0	20	–	–	–	–	–	–	–	–				
4167,	3	4	7	5	52	5 1/2	51 1/2	6373	0	4	4	0	22	0	22	–	–	–	–	–	–	–	–				
4265,	0	3	3	2	16	2 1/2	15 1/2	6781	1	3	4	14	24	16 3/8	21 5/8	–	–	–	–	–	–	–	–				
Brown,	0	1	1	0	3	0	3	–	–	–	–	–	–	–	–	–	–	–	–	–	–	–	–				
McLean,	0	4	4	1	13	0	14	–	–	–	–	–	–	–	–	–	–	–	–	–	–	–	–				

Several factorial explanations of the observed ratios between broodies and non-broodies in the several families can be developed, but choice among such explanations cannot be made because of the small size of the individual families, *i.e.*, the offspring of a single mother. Nor are any of them of value save as working hypotheses. The one on which Table II is based is presented simply to show that a close agreement between theory and fact is possible, and this theory was chosen for presentation because it gives a slightly better agreement between observed and theoretical rations, with one partial exception, than the others. This theory assumes that the appearance of broodiness in Rhode Island Reds requires the simultaneous presence of two factors, designated A and C, in either homozygous or heterozygous condition. A better fit in the case of the partial exception can be secured by assuming that there is also a dominant factor (presumably a modifier) for non-broodiness, which may be designated as N. Non-broodies, therefore, may be of numerous genetic types, the homozygous forms being NNAACC, nnAAcc, nnaaCC; and nnaacc, where A and C, respectively, represent the factors (condition of germ plasm) necessary for broodiness. A broody bird, then, in homozygous form must be nnAACC. As shown by Table III, which gives the theoreti-

TABLE III.— *The Theoretical Ratios resulting from Matings of Different Types, arranged by Ratios.*

[A ratio appears only once under its respective theory.]

Parents.	NNAACC Theory.		AACC Theory.	
	PROGENY.		PROGENY.	
	Non-broody.	Broody.	Non-broody.	Broody.
Broody × broody,	7	9	7	9
Non-broody × non-broody,	55 29 15 64	9 3 1 0	16	0
Broody × non-broody,	23	9	5	3
Broody × broody, broody × non-broody,	0 1	64 3	1	3
Broody × non-broody, non-broody × non-broody.	1 13 7	1 3 1	1 3	1 1
Broody × broody, broody × non-broody, non-broody × non-broody.	5 3	3 1	0	16

cal ratios expected on both the NNAACC and AACC theories, it will be seen that matings between birds of the same phænotype may give several different ratios, including those in which the proportions between broody and non-broody birds are reversed. Thus (NNAACC theory) broody × broody may give 3 broody to 1 non-broody, or it may give 1 broody to 3 non-broody, exactly reversing the ratio, as has occurred in the detailed data from which Table II was compiled. It should be stated that while families showing the extreme ratio of 15 non-broody to 1 broody have not been encountered, several instances of the 7:1 ratio have been observed.

The only evidence at present available in support of either of these schemes is furnished by the ratios between the broody and non-broody members of the several families (Table II). N, if it represents a real condition of the germ plasm, occurs relatively infrequently in the flock at present. Practically all the observed ratios, except the partial exception mentioned, can be accounted for if N is omitted.

It is also possible to modify the AACC theory, by assuming that A is sex linked, though no evidence of sex linkage other than an agreement between the observed and theoretical ratios has been noted. Doubtless other schemes could be devised that would also account for the ratios.

Although the ratios themselves could perhaps be explained as chance deviations from monohybrid ratios (though this is doubtful in some instances), or as the result of errors of classification of individuals through failure to manifest the genotypic condition phænotypically, the moment lines of descent are established it becomes clear that a monohybrid explanation does not fit the facts. The data have been worked over in an attempt to apply the monohybrid scheme, *i.e.*, broodiness due to a single dominant factor, but without success. See, for example, the history of male No. 3003 and his offspring, page 107.

In order to establish the existence of any of the schemes under discussion certain results of critical importance must be obtained. Thus, the discovery of a family consisting of all non-broody offspring from the mating between a non-broody and a broody is required to demonstrate the presence of a dominant factor for non-broodiness, while a mating between two non-broody birds that gives all broodies is required as proof of the AACC theory (or a theory of the same order). The ratios at hand indicate the possibility that several genetic types of non-broodies coexist in our strain. One possibility only seems to be excluded if the schemes outlined represent the facts, for one need never expect to find a pair of broody birds that produces all non-broody offspring, because such a result would mean two distinct types of broodies which mutually inhibit each other.

Modifying Factors for Broodiness.

The possibility that the non-broodies dealt with in these experiments are not due to changes in the primary genes concerned with broodiness, but are due to changes in modifying genes, cannot be excluded. As we have worked over the records, the impression has been strong that we are not dealing with a real absence of broodiness so much as with delay in the appearance of broodiness. Unfortunately the present data are inadequate to settle this point. Nor is it likely that we shall have suitable data in the near future, because the somatic manifestations of broodiness, *i.e.*, the number of times a bird becomes broody as well as the ease with which she is broken up, vary considerably, as already described. Since the chief reason for this variability is found in the number of times a bird becomes broody, which in turn is so thoroughly interwoven with egg production, the same practical difficulties, *i.e.*, disease control, that at present prevent a complete analysis of the inheritance of fecundity also prevent the determination of the hereditary factors involved in degree of broodiness.

The Production of a Strain of Low Degree of Broodiness through Selection.

Two lines of selection have been under way, — one for the elimination of broodiness, the other for its development to a high degree, equal to or greater than that observed in the case of Fig. 1. Because most of our facilities were needed in other directions, little has been done with the plus line beyond its maintenance. The minus line, however, has been closely involved with the problem of securing increased egg production, since absence of broodiness tends toward higher production, other things being equal. Until 1917 this line had been also carried on in a very small way, the general policy being to mate the son of a non-broody bird to non-broodies, on the hypothesis that broodiness is a simple Mendelian dominant, and non-broodiness a recessive. As a result of the early matings a male was obtained that appeared to be a homozygous recessive, since he threw no broodies from non-broody mothers. In 1917 this male, No. 3003, with his son, No. 5470 out of a non-broody hen, and grandson, No. 9752 (mother broody once in her third year), also supposed to be homozygous recessives, were mated to all the non-broody hens available. Some of these, however, became broody the second year. The results of the experiment, given in Tables IV and V, show that non-broodiness is not always a simple Mendelian recessive, since the son and grandson failed to breed true, even with those birds that never became broody. This

TABLE IV. — *The Progeny of Three Supposedly Non-broody Males distributed according to their Mother's Broody History.*

Male.	Mothers not Broody.			Mothers known to be Broody after Pullet Year.			Mothers Broody in Pullet Year.		
	Number of Mothers.	Daughters not Broody.	Daughters Broody.	Number of Mothers.	Daughters not Broody.	Daughters Broody.	Number of Mothers.	Daughters not Broody.	Daughters Broody.
No. 3003, ...	4	19	0	-	-	-	1	4	1
No. 5470 (son of No. 3003),	3	9	1	1	4	4	-	-	-
No. 9752 (son of No. 5470),	6	27	9	1	1	0	4	29	7

TABLE V. — "Non-broody" Lines, 1917–18.

[Daughters of males No. 3003, No. 5470 and No. 9752.]

	Number of Daughters.	DAUGHTERS NOT BROODY.		DAUGHTERS BROODY ONCE.		DAUGHTERS BROODY MORE THAN ONCE.		TOTAL BROODY DAUGHTERS.	
		Number.	Per Cent.	Number.	Per Cent.	Number.	Per Cent.	Number.	Per Cent.
Mothers not broody in pullet year.	72	58	80.55	9	12.50	5	6.94	14	19.44
Mothers broody once in pullet year.	34	28	82.35	1	2.94	5	14.71	6	17.65
Totals,	106	86	81.13	10	9.43	10	9.43	20	18.87

conclusion is supported by the ratios observed in other matings which have already been commented upon. However, the amount of broodiness in the first laying year is much reduced compared with the flock from which it originated, the data on this point being given in Table VII, Table VIII, item 4, and Table IX. A comparison with the published results of the laying contest at the Connecticut Agricultural Experiment Station shows that our foundation stock had broodiness developed to a higher degree than any of the breeds studied at Storrs, and that our non-broody

TABLE VI. — *Broodiness in the Several Breeds at the Storrs Contest of 1915–16, compared with Three Flocks at the Massachusetts Agricultural Experiment Station.*

BREEDS.	Number of Birds.	BROODY.		Average number of Times Broody per Broody Hen.	Average Number of Times Broody for All Birds in Flock.	Average Number of Days in Broody Period.	Average Number of Days spent in Broodiness by each Broody Hen.	Average Number of Days spent in Broodiness per Hen, per Year, all Birds included.
		Number.	Per Cent.					
Storrs.								
Plymouth Rocks,	151	67	44.4	2.8	1.2	21.2	59.9	26.6
Wyandottes,	151	87	57.6	2.5	1.4	19.4	47.6	27.4
Rhode Island Reds,	183	120	65.6	2.8	1.8	21.3	60.2	39.5
White Leghorns,	315	43	13.6	1.3	.2	22.7	29.6	4.0
Massachusetts.								
Rhode Island Reds, 1912–13.	125	112	89.6	4.4	3.9	19.7	74.8	65.8
Rhode Island Reds, 1913–14.	78	71	91.0	5.4	4.9	16.3	78.8	68.7
Rhode Island Reds, 1917–18, "non-broody" line.	106	20	18.9	1.9	.4	20.9	37.0	10.6

TABLE VII. — *Number and Per Cent of Birds Broody and not Broody in Pullet Year in Three Flocks of Rhode Island Reds.*

DATE.	Total Birds.	BROODY.		NOT BROODY.	
		Number.	Per Cent.	Number.	Per Cent.
1912–13, ...	125	112	89.60	13	10.40
1913–14, ...	78	71	91.03	7	8.97
1917–18, ...	106	20	18.87	86	81.13

lines, derived from this extremely broody stock, exhibited a low degree of broodiness surpassed only by the Leghorns (Table VI). The most significant data on this point are given in the third and last columns of Table VI. Table VII shows the relation between the number of broody birds and those not broody for the flocks of 1912–13, 1913–14 and 1917–18. The flock of 1912–13 was the foundation stock.

Tables VIII and X give a further comparison between the broody birds of the flocks of 1913–14 and 1917–18.

TABLE VIII. — *Statistical Constants for Various Broody Characters for the Flock of 1913–14, and the Non-broody Flock of 1917–18.*

		Number of Instances or Individuals.		Mean.		Standard Deviation.		Coefficient of Variation.	
		1913-14.	1917-18.	1913-14.	1917-18.	1913-14.	1917-18.	1913-14.	1917-18.
1	Number of broody periods per individual.	71	20	5.39±0.23	1.90±0.17	2.87±0.16	1.14±0.12	53.24±3.77	59.78±8.35
2	Length of broody periods (days),	327	34	16.28±0.34	20.91±0.95	9.04±0.24	8.22±0.67	55.60±1.85	39.32±3.68
3	Amount of broodiness in each individual (broodies only) (days).	68	20	78.84±4.03	37.00±3.63	49.27±2.85	24.06±2.57	71.68±5.90	65.02±9.42
4	Amount of broodiness in each individual (entire flock) (days)	78	106	68.73±3.28	10.63±1.08	42.95±2.32	16.46±0.76	62.49±4.50	154.83±17.27
5	Days in initial laying period,	71	20	118.67±4.23	170.50±8.42	52.82±2.99	55.86±5.96	44.51±2.98	32.76±3.85
6	Eggs in initial laying period,	71	20	80.64±2.85	101.50±3.84	35.66±2.02	25.48±2.72	44.22±2.95	25.10±2.84
7	Per cent production in initial laying period.	71	20	67.89±0.99	61.00±1.68	12.38±0.70	11.17±1.19	18.24±1.07	18.31±2.02
8	Days in each laying period,	327	34	18.84±0.40	36.85±2.98	10.85±0.29	25.80±2.11	57.60±1.96	70.00±8.06
9	Eggs in each laying period,	327	34	13.96±0.19	21.26±1.52	5.21±0.14	13.12±1.07	37.32±1.11	61.72±6.70
10	Per cent production in each laying period,	327	34	78.41±0.53	64.91±3.09	14.24±0.38	26.70±2.18	18.16±0.49	41.13±3.89
11	Time spent in laying periods annually by each individual (days).	68	20	90.75	62.75	—[1]	—[1]	—[1]	—[1]
12	Average length of broody cycle (days),	327	34	35.02±0.56	57.71±2.98	15.15±0.40	25.72±2.10	43.26±1.34	44.57±4.31
13	Per cent production in broody cycle,	327	34	41.50±0.43	35.94±1.54	11.55±0.30	13.29±1.09	27.83±0.79	36.98±3.41

[1] Standard deviation not calculated.   [2] Range 9–188.   [3] Range 16–114.

Definition of Terms used in Table VIII.

2. In reckoning the number of days in a broody period the first day without production is taken as the first broody day, while the last day counted is the day before production begins again. The object here is to measure the length of the non-productive period originating with broodiness, but not the intensity of broodiness itself. This definition includes instances in which the resumption of production is delayed long after its normal time because of the interference of factors not concerned with broodiness. Some limitation to the number of days included in the non-productive period is desirable, but the only one employed thus far is the exclusion of broody periods that end the annual cycle of production, and whose length cannot be ascertained.

5, 6, 7. The initial laying period begins with the first egg laid, and ends with the last egg laid before the first broody period.

8, 9, 10. A laying period begins with the first egg laid after a broody period, and ends with the last egg laid before the subsequent broody period. The incomplete cycles formed by a terminal broody period are rejected in calculating the constants. It is, of course, possible to treat the broody cycle somewhat differently, by defining it as a laying period plus the subsequent broody period.

12, 13. A broody cycle is defined as a broody period plus the following laying period. Biometrical constants were calculated for each method of treatment, but since the results proved to be essentially the same, if the initial cycle is omitted, only one set of constants is given in the table.

Constants differing slightly from those given in the table are obtained, if, instead of employing each instance separately in the calculations, the average for each individual bird is employed. Whether the instance or the average for each individual bird should be used in calculating the constants depends on which one occupies the center of interest, but whichever method is used, the primary purpose for which this table is presented is not affected. The inconsistencies in the number of individuals occur because it is often possible to determine a character in one individual but not in another. Thus, every bird that becomes broody can be counted, but if a bird becomes broody but once, and does not lay again until the following year, the length of her broody period cannot be measured, and so is omitted in calculating the constants.

Taking the means (Table VIII) as the basis of comparison, it is clear that the birds of the "non-broody" lines becoming broody in 1917–18 had the character much less intensely developed than the broodies of the flocks of 1912–13 and 1913–14 from which they originated. The mean number of times each broody bird became broody is 1.90 against 5.39. Though the average length of each broody period is longer (Table VIII), the total time spent in broodiness by each broody bird is about one-half that of the broodies of the flock of 1913–14. If the entire flocks of each year (*i.e.*, if the non-broody birds are included in calculating the means) are compared with each other the following significant results are obtained (Table IX):—

TABLE IX. — *A Comparison of the Amount of Broodiness in the Foundation Flock, 1912–13, and their Immediate Unselected Descendants, 1913–14, with their Descendants selected for the Absence of Broodiness, 1917–18.*

Date.	Number of Birds.	Mean Number of Days spent in Broodiness.	Mean Number of Times Broody.
1912–13,	125	65.81	3.88
1913–14,	78	68.73	4.91
1917–18,	106	10.63	.36

A comparison between the two flocks in respect to egg production (Table VIII) shows that while the 1917–18 flock laid somewhat less rapidly than the 1913–14 flock, the first broody period came later in life (Table X). The mean date of the first broody period is April 18 for the 1913–14 flock, and June 7 for the 1917–18 flock. The 1917–18 flock has a slower rate of production, as shown by the lower percentage production in the initial laying period as well as the later laying periods. On the other hand, the length, both of laying periods and broody periods, is longer

TABLE X. — *Seasonal Distribution of Broodiness in the Flock of 1913–14, and in the Broodies occurring in the Non-broody Lines, 1917–18.*

	Month.	Month in which Individual Broody Periods begin.				Month in which First Broody Period of Each Individual begins.				Month in which Last Broody Period of Each Individual begins.				Month in which Median Broody Period of Each Individual begins.			
		1913-14.		1917-18.		1913-14.		1917-18.		1913-14.		1917-18.		1913-14.		1917-18.	
		Number.	Per Cent.	Number.	Per Cent.	Number.	Per Cent.	Number.	Per Cent.	Number.	Per Cent.	Number.	Per Cent.	Number.	Per Cent.	Number.	Per Cent.
1	November,	–	–	–	–	–[1]	–[1]	–	–	–	–	–	–	–	–	–	–
2	December,	5	1.30	–	–	5	7.04	–	–	–	–	–	–	–	–	–	–
3	January,	4	1.04	–	–	3	4.23	–	–	–	–	–	–	–	–	–	–
4	February,	8	2.08	–	–	3	4.23	–	–	–	–	–	–	–	–	–	–
5	March,	19	4.94	–	–	10	14.08	–	–	–	–	–	–	–	–	–	–
6	April,	41	10.68	4	10.53	21	29.58	4	20.00	–	–	–	–	.5	.70	.5	2.50
7	May,	67	17.45	11	28.95	16	22.54	8	40.00	3	4.23	5	25.00	10.5	14.79	6.0	30.00
8	June,	64	16.67	6	15.79	9	12.68	2	10.00	4	5.63	3	15.00	34.0	47.89	5.0	25.00
9	July,	53	13.80	8	21.05	2	2.82	3	15.00	2	2.82	5	25.00	18.0	25.35	5.5	27.50
10	August,	48	12.50	4	10.53	–	–	3	15.00	6	8.45	2	10.00	5.0	7.04	2.0	10.00
11	September,	43	11.20	4	10.53	1	1.41	–	–	26	36.62	4	20.00	2.0	2.82	.5	2.50
12	October,	25	6.51	1	2.63	–	–	–	–	23	32.39	1	5.00	–	–	.5	2.50
13	November,	7	1.82	–	–	1	1.41	–	–	7	9.86	–	–	1.0	1.41	–	–
	Totals,	384	99.99	38	100.01	71	100.02	20	100.00	71	100.00	20	100.00	71.0	100.00	20.0	100.00

[1] Very few birds laying.

in this flock than in that of 1913–14. Just what this means is uncertain. The longer laying periods may be taken as resultant of the reduced tendency toward broodiness, but this is not true for the longer broody periods. The latter may be connected with the slower rate of production.

The experiments in eliminating broodiness are being continued, but a change in the plan of the experiment, to permit of the fusion of the non-broody line with another line known as the high-producing line, has been made. The fusion appears at date of this writing to be accomplished.

Broodiness, in its various sub-characters and in the associated periods of egg production, is decidedly variable as judged by the several coefficients of variation given in Table VIII. Some of the sub-characters are much more variable than others. While some of the characters associated with broodiness are of the same order of variability in the two flocks studied, others are quite unlike, sometimes one and sometimes the other flock being the more variable. The details are best obtained from Table VIII.

Relation between Birds of a Low Degree of Broodiness and Absence of Broodiness.

Some evidence exists that birds that become broody once during the pullet year are not genetically different from those that do not become broody, since the number of broody offspring from each sort of female is approximately the same, as is shown in Table XI. On the other hand,

Table XI. — *A Comparison between the Number of Broody Offspring from Non-broody Mothers with the Number from Mothers Broody once, the Sires being the Same for Both Lots of Offspring.*

	Number of Mothers.	Broody Offspring.		Non-broody Offspring.	
		Number.	Per Cent.	Number.	Per Cent.
Not broody, ...	15	14	19.45	58	80.55
Broody, ...	3	6	17.65	28	82.35

the daughters of birds broody once are somewhat more broody than the daughters of birds not broody at all, as shown in Table XII, which gives a comparison between 14 broody daughters of non-broody mothers and 6 broody daughters of mothers that became broody once, the sires being the same for both lots. It is shown by the per cent production, for both the initial laying period and the subsequent laying periods, that the two sets of birds are about equal in their ability to produce eggs. The daughters whose mothers became broody once were, however, somewhat more broody than the daughters of hens that did not become broody at all, as shown by the length of the initial laying period, the number of broody periods per individual, and the length of the broody periods. Though in this experiment the daughters of non-broody hens are less broody than the daughters of hens broody once, it would be unwise to generalize such a conclusion, because of the very small number of individuals involved.

TABLE XII. — *A Comparison of the Amount of Broodiness in the Daughters of Non-broody Hens with those whose Mothers became Broody once.*

	MOTHERS NOT BROODY IN PULLET YEAR.		MOTHERS BROODY ONCE IN PULLET YEAR.	
	Number of Instances or Individuals.	Mean.	Number of Instances or Individuals.	Mean.
Days of broodiness per individual,	14	31.64	6	49.50
Days in each broody period,	22	18.26	12	24.25
Broody periods per individual,	14	1.79	6	2.17
Days in initial laying period,	14	181.36	6	145.67
Eggs laid in initial laying period,	14	106.93	6	85.83
Per cent production in initial laying period per individual.	14	60.86	6	60.27
Days in each laying period,	22	38.23	12	34.50
Eggs laid in each laying period,	22	21.95	12	19.83
Per cent production in each laying period,	22	65.20	12	64.70
Eggs in each laying period per individual,	14	25.44	6	21.37

Since some birds become broody in their second or third laying years that did not become broody in the first year, the question may be raised as to whether or not a hen may ever be so constituted that it is impossible for her to become broody. We have kept a few hens for four years without evidence of broodiness, but this may not mean that these birds might not have become broody if the proper stimulus had existed. There is the further question as to whether the designation "non-broody" has been accurately used for birds not broody in their pullet year. It might be better to regard such cases as instances of delayed broodiness rather than of the actual absence of broodiness. The delay in the appearance of broodiness in some individuals certainly complicates matters greatly.

THE INTERRELATION OF SEVERAL BROODY CHARACTERS.

The interrelations of several of the broody characters have been studied in the 1913-14 flock by means of the coefficient of correlation. It should, perhaps, be pointed out that the coefficient of correlation does not measure the relationship between the characters as such, but relationship between the numerical occurrence of such characters in the flock studied. This limitation in the use of the coefficient of correlation is often forgotten. Thus it is found that r between number of eggs laid in a year and total days spent in broodiness is $+.1677 \pm .0742$. This value, as shown by its large probable error, is not significant statistically, but, ignoring the error, may perhaps indicate that broodiness is an advantage, since, on the average, those birds spending the most time in broodiness are the heaviest layers. On the contrary, it is known from a study of other data that the very best layers cannot spend much time in broodiness. The interpretation we give this value is that those birds whose laying year begins earliest and stops latest get in more broody periods, other things being equal, than birds whose laying year is shorter.

If an index of production of high value is desired, it is found in the initial laying period, for here the correlation between the length of the period and number of eggs produced is very high, viz., $+.8843 \pm .0210$, a value, moreover, that indicates good homogeneity in rate of production in this flock.

In this flock there is a pronounced negative correlation between egg production during the laying periods and number of broody periods, the coefficient of correlation being $-.3453 \pm .0716$, indicating that those birds that are very broody tend to lay less eggs between broody periods than those having a less number of broody periods. On the other hand, there is no relation between the average (i.e., for one individual) length of laying periods or the eggs produced in such periods and average length of broody periods, since in the first case $r = -.0130 \pm .0818$, and in the second case $r = -.0013 \pm .0818$.

While the above statements hold true for average values, if the coefficient of correlation is determined between the length of a laying period or its egg production and the length of the broody period immediately subsequent thereto, a marked negative correlation is found, being $-.2899 \pm .0415$ in the first instance, and $-.3715 \pm .0345$ in the second. The disagreement between the values obtained when each laying period is correlated with its subsequent broody period, and that found when the average value for each bird is used, is due to a shortening of the laying period and a lengthening of the broody period as the season progresses. This is clearly shown on the individual records.

If, instead of taking a laying period and its subsequent broody period, a broody period is paired with the laying period following, little or no relationship is indicated, for r between length of broody period and subsequent laying period is $-.0222 \pm .0388$, while between length of broody period and subsequent egg production it is only slightly greater, being $-.0799 \pm .0372$.

The interrelationships discussed in the two paragraphs preceding may perhaps be interpreted to mean that heavy laying tends to suppress broodiness, or, at least, that in the flock studied, those birds that laid most heavily had shorter broody periods than those laying less heavily, the tendency to heavy production in such birds enabling them to get back more quickly into production than those in which the tendency was less strong. Longer broody periods, however, and their accompanying element of rest did not conduce to heavier production, a view contrary to that held by most poultrymen.

Summary and Conclusions.

The working hypothesis is adopted that —

1. Broodiness depends upon the presence of a "complete mechanism" in the individual, from which it follows that the absence of broodiness depends upon the loss of some essential part of this mechanism, or upon its inhibition by some secondary factor.

2. The inheritance of broodiness may be expected to vary from flock to flock.

3. In the flocks studied, non-broodiness appears to result from the loss of one or both of two genes from the complete germinal complex, while there is some evidence that a dominant inhibitor may also exist in the germ plasm of these flocks.

4. By suitable breeding methods it has been possible to develop quickly a strain of low degree of broodiness from a strain with a very high degree of broodiness.

5. Statistical constants for certain broody characters are given.

Literature Cited.

Bateson, W., 1902. Experiments with Poultry. Reports to the Evolution Committee of the Royal Society. Report I, pp. 87–124.

Gerhartz, H., 1904. Ueber die zum Aufbau der Eizelle Notwendige Energie. (Transformationsenergie) Pfluger's Arch. Bd. 56, pp. 1–224.

Goodale, H. D., 1918. Internal Factors influencing Egg Production in the Rhode Island Red Breed of Domestic Fowl. Am. Nat, Vol. LII, pp. 65–94, 209–232, 301–321.

Herrick, F. H., 1907a. Analysis of the Cyclical Instincts of Birds. Science, N. S., Vol. XXV, pp. 725, 726.

———, 1907b. The Blending and Overlap of Instincts. Science, N. S., Vol. XXV, pp. 781, 782.

Hurst, C. C., 1905. Experiments with Poultry. Reports to the Evolution Committee of the Royal Society. Report II, pp. 131–154.

Kirkpatrick, W. F., and Card, L. E., 1917. Fifth Annual International Egg-laying Contest. Bulletin No. 89, Storrs Agricultural Experiment Station, pp. 257–301.

Pearl, R., 1914. Studies on the Physiology of Reproduction in the Domestic Fowl. VII. Data regarding the Brooding Instinct in its Relation to Egg Production. Journal An. Beh., Vol. 4, pp. 266–288.

www.ingramcontent.com/pod-product-compliance
Lightning Source LLC
Chambersburg PA
CBHW062345220526

45469CB00008B/2842